Name _____

Number Tracks

Write the numbers.

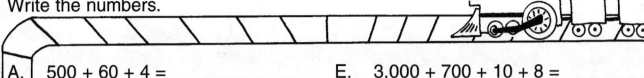

A. 500 + 60 + 4 = _____

B. 300 + 20 + 9 = _____

C. 800 + 3 = _____

D. 200 + 70 = _____

E. 3,000 + 700 + 10 + 8 = _____

F. 5,000 + 800 + 20 + 5 = _____

G. 4,000 + 50 + 2 = _____

H. 9,000 + 400 + 9 = _____

Look for a number pattern. Write the missing numbers.

I. 78, 79, 80, _____, _____, 83, 84, _____, 86

J. 437, 438, _____, _____, 441, 442, 443, _____

K. 2,109; 2,110; 2,111; _____; _____; 2,114; _____

L. 320, 325, 330, _____, _____, _____, 350

M. 4,560; 4,562; 4,564; _____; _____; _____

Find the mystery number.

N. I have a 3 in my ones place. My tens digit is 4 more than my hundreds digit. My hundreds digit is 2 more than my ones digit. Who am I?

O. I have a 2 in my tens place. My hundreds digit is 3 less than my ones digit. My ones digit is 2 more than my tens digit. Who am I?

MP3493

Name _____

Space Facts

Add or subtract.

A.
$$\begin{array}{r} 3 \\ + 5 \\ \hline \end{array}$$
$$\begin{array}{r} 7 \\ + 6 \\ \hline \end{array}$$
$$\begin{array}{r} 4 \\ + 9 \\ \hline \end{array}$$
$$\begin{array}{r} 8 \\ + 2 \\ \hline \end{array}$$
$$\begin{array}{r} 2 \\ + 3 \\ \hline \end{array}$$
$$\begin{array}{r} 7 \\ + 7 \\ \hline \end{array}$$
$$\begin{array}{r} 1 \\ + 8 \\ \hline \end{array}$$

B.
$$\begin{array}{r} 9 \\ - 3 \\ \hline \end{array}$$
$$\begin{array}{r} 11 \\ - 7 \\ \hline \end{array}$$
$$\begin{array}{r} 7 \\ - 4 \\ \hline \end{array}$$
$$\begin{array}{r} 13 \\ - 5 \\ \hline \end{array}$$
$$\begin{array}{r} 10 \\ - 4 \\ \hline \end{array}$$
$$\begin{array}{r} 12 \\ - 6 \\ \hline \end{array}$$
$$\begin{array}{r} 15 \\ - 7 \\ \hline \end{array}$$

C.
$$\begin{array}{r} 5 \\ + 2 \\ \hline \end{array}$$
$$\begin{array}{r} 9 \\ + 5 \\ \hline \end{array}$$
$$\begin{array}{r} 8 \\ + 6 \\ \hline \end{array}$$
$$\begin{array}{r} 4 \\ + 8 \\ \hline \end{array}$$
$$\begin{array}{r} 7 \\ + 5 \\ \hline \end{array}$$
$$\begin{array}{r} 5 \\ + 6 \\ \hline \end{array}$$
$$\begin{array}{r} 5 \\ + 8 \\ \hline \end{array}$$

D.
$$\begin{array}{r} 16 \\ - 8 \\ \hline \end{array}$$
$$\begin{array}{r} 10 \\ - 2 \\ \hline \end{array}$$
$$\begin{array}{r} 12 \\ - 9 \\ \hline \end{array}$$
$$\begin{array}{r} 17 \\ - 8 \\ \hline \end{array}$$
$$\begin{array}{r} 14 \\ - 5 \\ \hline \end{array}$$
$$\begin{array}{r} 11 \\ - 3 \\ \hline \end{array}$$
$$\begin{array}{r} 15 \\ - 6 \\ \hline \end{array}$$

E.
$$\begin{array}{r} 6 \\ + 6 \\ \hline \end{array}$$
$$\begin{array}{r} 18 \\ - 9 \\ \hline \end{array}$$
$$\begin{array}{r} 4 \\ + 2 \\ \hline \end{array}$$
$$\begin{array}{r} 11 \\ - 9 \\ \hline \end{array}$$
$$\begin{array}{r} 5 \\ + 5 \\ \hline \end{array}$$
$$\begin{array}{r} 15 \\ - 8 \\ \hline \end{array}$$
$$\begin{array}{r} 8 \\ + 9 \\ \hline \end{array}$$

F.
$$\begin{array}{r} 2 \\ + 5 \\ \hline \end{array}$$
$$\begin{array}{r} 10 \\ - 9 \\ \hline \end{array}$$
$$\begin{array}{r} 9 \\ + 9 \\ \hline \end{array}$$
$$\begin{array}{r} 12 \\ - 2 \\ \hline \end{array}$$
$$\begin{array}{r} 6 \\ + 9 \\ \hline \end{array}$$
$$\begin{array}{r} 10 \\ - 3 \\ \hline \end{array}$$
$$\begin{array}{r} 8 \\ + 8 \\ \hline \end{array}$$

Name _____

Pet Palace

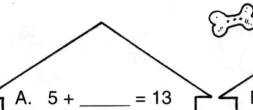

Write the missing numbers.

A. 5 + _____ = 13

8 + _____ = 13

13 − _____ = 5

_____ − 5 = 8

B. 7 + 9 = _____

9 + _____ = 16

_____ − 9 = 7

16 − _____ = 9

C. _____ + 10 = 19

10 + _____ = 19

19 − _____ = 9

_____ − 9 = 10

D. _____ + 8 = 11

8 + _____ = 11

_____ − 8 = 3

11 − _____ = 8

E. 6 + _____ = 15

_____ + 6 = 15

15 − 9 = _____

_____ − 6 = 9

F. 10 + _____ = 17

_____ + 10 = 17

17 − _____ = 10

_____ − 10 = 7

G. 4 + _____ = 12

_____ + 4 = 12

_____ − 8 = 4

12 − _____ = 8

H. 8 + 6 = _____

6 + _____ = 14

14 − _____ = 8

_____ − 8 = 6

I. 7 + _____ = 13

_____ + 7 = 13

13 − _____ = 7

_____ − 7 = 6

         MP3493

Fruit Basket Upset

Add. Remember to regroup if you need to.

A.	23 + 14	35 + 21	50 + 18	17 + 61	45 + 30	15 + 22
B.	46 + 27	61 + 28	17 + 41	14 + 32	33 + 29	5 + 13
C.	47 + 9	35 + 51	42 +52	48 + 17	70 + 7	18 + 34
D.	14 + 56	54 + 9	30 + 87	98 + 33	49 + 5	86 + 70
E.	91 + 37	58 + 91	80 + 56	37 + 88	59 + 51	73 + 94

Find the missing addend. Hint: Answers from row C will solve each problem.

F.

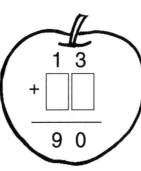
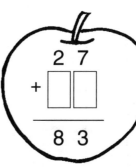
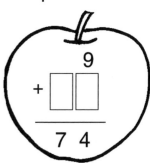

Name _____

Subtraction Action

Subtract. Remember to regroup if you need to.

A.
```
  57        48        92        96        75        83
- 24      - 16      - 50      - 32      - 24      - 43
```

B.
```
  52        73        68        95        83        34
- 37      - 21      - 29      - 31      - 15      - 13
```

C.
```
  56        67        85        90        71        35
-  8      - 50      - 42      - 67      -  6      - 27
```

D.
```
  99        54        72        81        76        92
- 28      - 16      - 63      - 33      - 44      - 20
```

E.
```
  56        77        85        44        69        38
- 37      - 19      - 55      - 18      - 51      - 21
```

Subtract down, then across. Find the secret number in the box.

F.

8	3	
4	2	

G.

9	3	
3	2	

H.

7	2	
6	1	

                   MP3493

Name _____

Bean Bag Fun

Add. Remember to regroup if you need to.

A.   561   156   329   397   866
    + 210   + 243   + 208   + 424   + 415

B.   749   515   246   358   257
    + 132   + 48   + 133   + 109   + 614

C.   352   373   189   716   567
    + 173   + 445   + 304   + 59   + 426

D.   467   289   412   167   345
    + 266   + 134   + 198   + 656   + 296

E.   734   679   473   759   845
    + 127   + 415   + 367   + 589   + 490

Marta and Paul play bean bag. What two numbers do they need to land on?

Marta needs to score 300 points.   Paul needs to score 400 points.

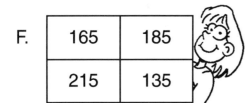

F.

165	185
215	135

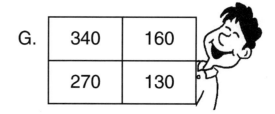

G.

340	160
270	130

Marta_____   Paul_____

Tic Tac Numbers

Subtract. Remember to regroup if you need to.

A.	645 − 134	963 − 251	682 − 410	845 − 34	736 − 415
B.	487 − 149	753 − 226	281 − 139	540 − 107	956 − 99
C.	819 − 162	549 − 94	278 − 193	487 − 195	947 − 262
D.	677 − 189	945 − 687	856 − 278	933 − 65	718 − 239
E.	873 − 172	584 − 297	874 − 609	566 − 180	831 − 176

Subtract down to find the missing numbers.

F.

	5	9
4		8
1	3	

G.

7	6	
4		3
	2	5

H.

	8	9	9
2	6	8

MP3493

Name _____

Farmer's Market

Add or subtract.

A.
```
    67        92        45        32        86
  + 25      − 38      + 67      + 58      − 59
```

B.
```
   348       436       805       487       826
  − 215      + 94      − 92      + 434     − 187
```

C.
```
   415       316       600       361       315
  + 162      + 589     − 318     − 127     + 674
```

Mr. Mott counted the bushels of vegetables he had at the Farmer's Market on Monday. He wrote the information in a chart. Help Mr. Mott keep track of his vegetables on Tuesday, Wednesday, and Thursday.

Vegetable	Bushels on Monday
Carrots	65
Lettuce	23
Spinach	47
String Beans	105
Tomatoes	52

D. On Tuesday, Mr. Mott gathers 49 more bushels of carrots. How many bushels does he have now?

_____

E. On Wednesday, Mr. Mott sells 26 bushels of string beans. How many bushels are left?

_____

F. On Thursday, Mr. Mott sells 12 bushels of tomatoes and 10 bushels of spinach. How many bushels of tomatoes and spinach does he have now?

_____

MP3493

Get a Clue!

Add.

A.	1,476 + 2,563	2,805 + 4,127	6,328 + 439	3,211 + 2,503	1,895 + 2,342
	O	**M**	**K**	**O**	**A**

B.	4,065 + 3,872	1,094 + 784	4,321 + 2,093	635 + 6,732	8,005 + 1,564
	A	**R**	**C**	**D**	**Q**

C.	1,069 + 1,484	2,515 + 3,197	4,673 + 3,648	43,211 + 2,503	61,895 + 2,342
	F	**S**	**Y**	**B**	**P**

D.	13,045 + 2,965	41,908 + 35,674	45,226 + 31,682	11,944 + 24,503	81,895 + 12,342
	U	**E**	**G**	**X**	**H**

E. What do you get when you put a bunch of ducks in a box? To solve the riddle, match the letters from your answers above to the numbers below the lines.

_____		_____	_____	_____		_____	_____
4,237		45,714	4,039	36,447		5,714	2,553

_____	_____	_____	_____	_____	_____	_____	_____
9,569	16,010	7,937	6,414	6,767	77,582	1,878	5,712

Name _____

Magic Numbers

Subtract.

A.	4,683	6,116	3,485	9,005	8,538
	− 2,563	− 1,183	− 873	− 2,456	− 3,056

B.	7,483	2,637	8,456	4,153	6,894
	− 3,384	− 924	− 2,999	− 1,148	− 5,975

C.	1,885	3,843	9,343	34,890	55,078
	− 697	− 3,457	− 2,889	− 5,211	− 4,119

D.	44,000	88,341	55,839	70,670	91,059
	− 5,576	− 42,875	− 49,415	− 18,345	− 14,328

Extra Challenge! Subtract down to find the magic numbers. Write them in the hats.

E.

7	3	2	11	5
4	0	1	2	
		1	9	2

F.

	7	8	11	9
3	2	6		4
2		2	6	

The magic number is

The magic number is

MP3493

At the Arcade

Use the game boards to solve each word problem. The boards show the number of points scored by each player.

SPACE RAIDERS		AMAZON TREK		PIRATE'S ISLAND	
Meagan	389	Latrell	4,189	Nina	32,977
Tom	857	Misha	6,037	Jamal	45,062
Raul	518	Nicky	7,834	Lin	52,944

A. How many points did Meagan and Raul score in Space Raiders? Together did they beat Tom's score?

_____

B. Latrell scores 3,671 more points playing Amazon Trek. Did he score enough total points to beat Nicky?

_____

C. Misha wants to find out how many more points he needs to tie scores with Nicky. How can he find out? How many points does he need?

_____

D. Which two players scored the most points playing Pirate's Island? How many total points did these two players score?

_____

E. How many more points did the highest scoring player in Pirate's Island score than the lowest scoring player?

_____

F. Jamal scores 10,785 more points playing Pirate's Island. What is his new total? Is it enough to beat Lin's score?

_____

Play Ball!

Multiply by 2.

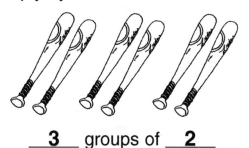

_____**3**_____ groups of _____**2**_____

3 x 2 = _**6**_

Multiply by 5.

_____**3**_____ groups of _____**5**_____

3 x 5 = _**15**_

Multiply.

A.   5 x 5 = _____   2 x 2 = _____   4 x 2 = _____

B.   8 x 2 = _____   1 x 5 = _____   9 x 5 = _____

C.

2	5	2	5	2	5	2
x 6	x 3	x 5	x 4	x 9	x 6	x 3

D.

5	2	8	7	1	3	5
x 2	x 7	x 5	x 5	x 2	x 5	x 9

E.

2	6	9	6	0	4	7
x 8	x 5	x 2	x 2	x 5	x 5	x 2

F. Write a multiplication sentence for 5 + 5 + 5 + 5 + 5. _____

Crazy Cats

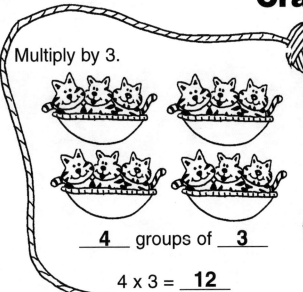

Multiply by 3.

4 groups of **3**

4 x 3 = **12**

Multiply by 4.

5 groups of **4**

5 x 4 = **20**

Multiply.

A.   1 x 3 = _____   2 x 3 = _____   7 x 4 = _____

B.   5 x 4 = _____   3 x 3 = _____   8 x 4 = _____

C.
4	5	2	2	4	1	7
x 6	x 3	x 4	x 3	x 2	x 4	x 3

D.
4	4	0	3	6	3	3
x 3	x 7	x 4	x 9	x 3	x 4	x 5

E.
3	9	9	6	8	4	3
x 8	x 3	x 4	x 4	x 3	x 4	x 6

F.   Write a multiplication sentence for 8 + 8 + 8 + 8. _____

                                         MP3493

Name _____

It's Fruity!

Count the groups. Multiply.

3 x 6 = 18

3 x 7 = 21

3 x 8 = 24

2 x 9 = 18

A.   6 x 7 = _____   7 x 7 = _____   8 x 6 = _____

B.   9 x 6 = _____   5 x 7 = _____   7 x 8 = _____

C.
9	6	7	5	5	7	4
x 7	x 6	x 9	x 8	x 9	x 6	x 6

D.
7	8	6	8	9	6	8
x 4	x 8	x 8	x 9	x 9	x 9	x 7

E.   Find the missing numbers.

_____ x 7 = 49   6 x _____ = 48   _____ x _____ = 64

14   MP3493

Hanging Around

Multiply. Then color the squares that have answers ending in 0 or 5.

A. 6 x 7 **B**	5 x 4 **T**	9 x 3 **E**	8 x 1 **N**	4 x 3 **Y**	2 x 7 **T**	5 x 5 **A**
B. 3 x 7 **V**	8 x 8 **G**	5 x 7 **F**	9 x 2 **P**	6 x 4 **O**	8 x 3 **M**	9 x 7 **K**
C. 6 x 5 **O**	4 x 4 **Q**	6 x 9 **R**	8 x 5 **C**	3 x 3 **W**	6 x 8 **Y**	7 x 7 **J**
D. 9 x 9 **H**	4 x 7 **Z**	8 x 7 **X**	6 x 3 **A**	4 x 8 **D**	5 x 3 **R**	6 x 6 **S**

E. What kind of tree has numbers in it that are multiplied? To find out, unscramble the letters in the squares you colored above.

A _____ tree

F. Write numbers in the top row that complete each multiplication fact. The numbers in the middle and bottom rows are the products.

x	3			
4	12	8	20	28
9	27	18	45	63

MP3493

Name _____

Bird Watching

Divide by 2.

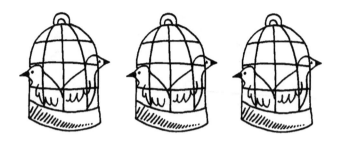

_____**2**_____ in each group

$6 \div 2 =$ _____**3**_____

Divide by 5.

_____**5**_____ in each group

$15 \div 5 =$ _____**3**_____

Divide.

A. $8 \div 2 =$ _____ $10 \div 5 =$ _____ $12 \div 2 =$ _____ $4 \div 2 =$ _____

B. $16 \div 2 =$ _____ $18 \div 2 =$ _____ $30 \div 5 =$ _____ $45 \div 5 =$ _____

C. $5\overline{)15}$ $2\overline{)16}$ $2\overline{)12}$ $2\overline{)14}$ $2\overline{)6}$

D. $5\overline{)10}$ $5\overline{)35}$ $5\overline{)20}$ $5\overline{)5}$ $2\overline{)10}$

E. $5\overline{)25}$ $2\overline{)2}$ $5\overline{)40}$ $2\overline{)20}$ $5\overline{)35}$

F.   How many times can you subtract 5 from 50? _____

Name _____

Art Class

Divide by 3.

___3___ in each group

$6 \div 3 =$ ___2___

Divide by 4.

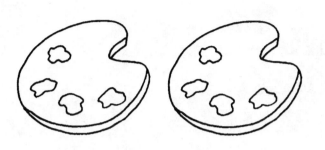

___4___ in each group

$8 \div 4 =$ ___2___

Divide.

A. $9 \div 3 =$ _____ $4 \div 4 =$ _____ $28 \div 4 =$ _____ $24 \div 3 =$ _____

B. $12 \div 4 =$ _____ $8 \div 4 =$ _____ $32 \div 4 =$ _____ $20 \div 4 =$ _____

C. $3\overline{)18}$ $3\overline{)21}$ $4\overline{)16}$ $3\overline{)3}$ $4\overline{)24}$

D. $3\overline{)27}$ $4\overline{)40}$ $4\overline{)36}$ $4\overline{)12}$ $3\overline{)15}$

E. $3\overline{)6}$ $4\overline{)20}$ $3\overline{)30}$ $3\overline{)12}$ $4\overline{)28}$

F. How many times can you subtract 4 from 32? _____

Name _____

Let's Play Chess

Think of a multiplication fact to help you divide.

 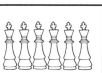	6 x 6 = __36__ 36 ÷ 6 = __6__

5 x 7 = 35 35 ÷ 7 = __5__	3 x 8 = 24 24 ÷ 8 = __3__	4 x 9 = 36 36 ÷ 9 = __4__

Divide.

A. 24 ÷ 6 = _____   49 ÷ 7 = _____   64 ÷ 8 = _____   30 ÷ 6 = _____

B. 56 ÷ 7 = _____   63 ÷ 9 = _____   81 ÷ 9 = _____   42 ÷ 6 = _____

C. $8\overline{)40}$ $8\overline{)72}$ $6\overline{)48}$ $7\overline{)63}$ $6\overline{)36}$

D. $8\overline{)56}$ $6\overline{)54}$ $7\overline{)42}$ $6\overline{)18}$ $7\overline{)28}$

E. $9\overline{)54}$ $8\overline{)48}$ $9\overline{)72}$ $9\overline{)45}$ $8\overline{)32}$

A Fishy Story

Divide. Then color the squares that have 3 or 9 as their answers.

A.						
$4\overline{)32}$	$8\overline{)16}$	$9\overline{)18}$	$3\overline{)18}$	$4\overline{)16}$	$4\overline{)28}$	$9\overline{)81}$
V	F	X	A	S	M	O
B.						
$5\overline{)40}$	$6\overline{)30}$	$4\overline{)12}$	$8\overline{)56}$	$7\overline{)21}$	$5\overline{)30}$	$3\overline{)24}$
K	B	U	F	G	D	H
C.						
$9\overline{)72}$	$6\overline{)54}$	$7\overline{)42}$	$9\overline{)27}$	$6\overline{)24}$	$9\overline{)9}$	$2\overline{)14}$
Y	P	T	R	Z	L	P
D.						
$5\overline{)20}$	$7\overline{)49}$	$9\overline{)36}$	$8\overline{)24}$	$8\overline{)64}$	$7\overline{)63}$	$6\overline{)48}$
Q	N	J	R	W	E	C

E. Unscramble the letters in the colored squares above to solve this riddle:
What kind of fish likes to divide?

A _____ fish

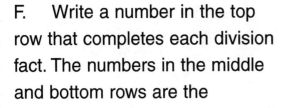

F. Write a number in the top row that completes each division fact. The numbers in the middle and bottom rows are the quotients.

÷	24			
3	8	4	6	10
6	4	2	3	5

   MP3493

Hat Tricks

Multiplying and dividing with 1.

$1 \times 3 =$ __3__

$3 \times 1 =$ __3__

$3 \div 1 =$ __3__

Multiplying and dividing with 0.

$0 \times 3 =$ __0__

$3 \times 0 =$ __0__

$0 \div 3 =$ __0__

Multiply.

A. $1 \times 7 =$ _____ $5 \times 0 =$ _____ $8 \times 1 =$ _____

B. $9 \times 1 =$ _____ $0 \times 4 =$ _____ $0 \times 8 =$ _____

C.
9	1	7	0	6	5	1
x 0	x 6	x 1	x 2	x 1	x 1	x 0

Divide.

D. $6 \div 1 =$ _____ $0 \div 7 =$ _____ $4 \div 1 =$ _____ $0 \div 8 =$ _____

E. $0 \div 5 =$ _____ $9 \div 1 =$ _____ $8 \div 1 =$ _____ $1 \div 1 =$ _____

F. $1\overline{)4}$ $1\overline{)7}$ $8\overline{)0}$ $6\overline{)0}$ $1\overline{)3}$

Find the missing numbers.

G. $1 \times \boxed{} = 8$ $\boxed{} \times 6 = 0$ $7 \times \boxed{} = 0$ $\boxed{} \div 1 = 2$

                   MP3493

Name _____

Off to the Races!

Multiply and divide.

A.
9	2	7	8	6	5	3
x 3	x 6	x 4	x 2	x 5	x 7	x 3

B.
7	3	4	8	6	3	5
x 9	x 7	x 4	x 7	x 8	x 4	x 5

C.
6	8	5	8	7	9	6
x 4	x 3	x 9	x 4	x 7	x 9	x 9

D. 4)20   9)36   8)64   6)12   5)15

E. 1)6   8)56   3)15   9)0   7)21

F. 4)8   6)36   8)40   9)9   7)42

G. Which car crosses the finish line first? To find out, color the numbers in the boxes below that match the products and quotients above.

Products (x)	49	24	27	54	35	16	36
Quotients (÷)	4	8	5	2	1	6	3

Car _____ finishes first.

MP3493

Name _____

At the State Fair

Write a number sentence to solve each problem.

A. Maya bakes cupcakes for the fair. She puts 8 cupcakes in a tray. She bakes 7 trays. How many cupcakes is that?

_____

B. There is room for 45 riders on the roller coaster. The roller coaster has 9 cars. How many riders can fit in each car?

_____

C. Jamal has 36 balls at the ball toss. He puts them in 6 stacks. How many balls are in each stack?

_____

D. A group of 35 people go to the State Fair. They ride in 5 vans. How many people can ride in each van?

_____

E. There are 5 prize-winning pigs in each pen. There are 4 pens in all. How many prize-winning pigs are there?

_____

F. Tickets come in booklets of 5. The O'Conner family buys 6 booklets. How many tickets is that in all?

_____

G. Mr. Mooney has 8 boxes of apple pies. Each box contains 4 pies. How many pies in all does Mr. Mooney have at his stand?

_____

H. The chicken coop has 7 cages for the chickens. Each cage holds 4 chickens. How many chickens are there in all?

_____

Name _____

Flat Land

A. Color a box on the graph for each flat shape.

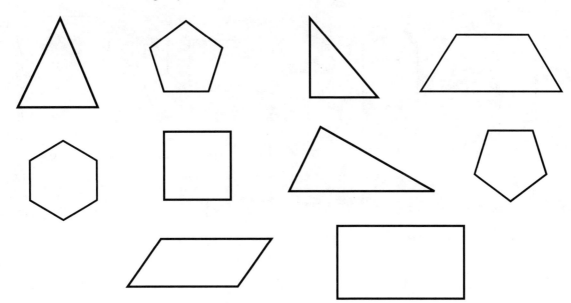

Flat Shapes

	0	1	2	3	4	5	6	7
Triangle	▓	▓	▓					
Quadrilateral								
Pentagon								
Hexagon								

Number of Shapes

Use the information in the graph to write a number sentence for each problem.

B. △ + ⬡ = __?__   C. ☐ + ⬠ = __?__

___ + ___ = ___   ___ + ___ = ___

23   MP3493

Name _____

A Solid State

A.   Color a box on the graph for each solid shape.

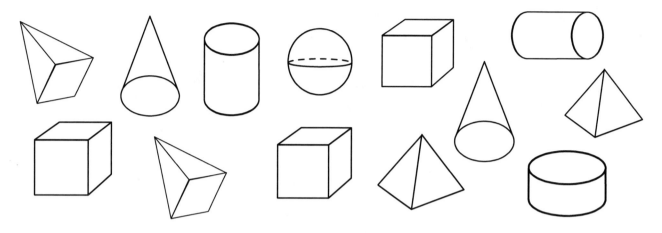

Solid Shapes

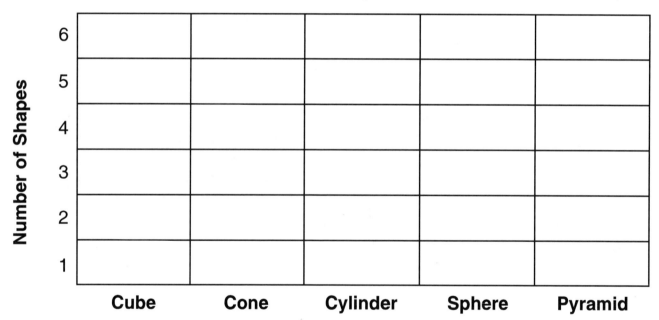

Use the graph to write a number sentence for each problem.

B. △ + ⬜ + △ = __?__   C. ⬭ + ⬜ = __?__

____ + ____ + ____ = ____   ____ + ____ = ____

                                       MP3493

Name _____

Shape Parts

Some solid shapes have 3 parts.

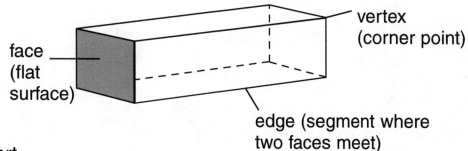

face
(flat
surface)

vertex
(corner point)

edge (segment where
two faces meet)

Complete the chart.

Solid Shape	Shape of Flat Faces	Number of Flat Faces	Number of Straight Edges	Number of Vertices
A. Cube				
B. Pyramid				
C. Cone				
D. Cylinder				
E. Sphere				

Extra Challenge! What two solids make up this picture?

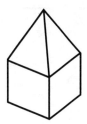

_____

Pie Time

Color each fraction.

A. Color $\frac{1}{3}$

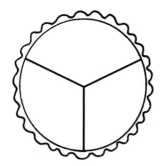

B. Color $\frac{2}{5}$

C. Color $\frac{3}{4}$

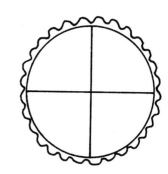

Write the fraction for the part that is shaded.

D.

E.

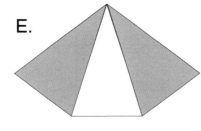

F.

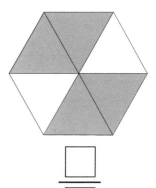

G. Make deluxe pizza. Draw mushrooms on $\frac{3}{8}$ of the pie. Draw sausage on $\frac{5}{8}$ of the pie.

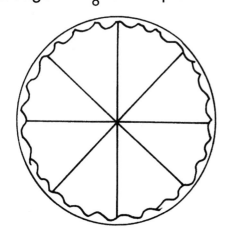

H. Make vegetable pizza. Draw peppers on $\frac{2}{6}$ of the pie. Draw broccoli on $\frac{4}{6}$ of the pie.

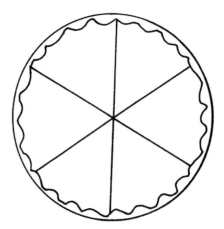

At the Craft Shop

Color the fraction of the group.

A. Color $\frac{1}{4}$

B. Color $\frac{4}{6}$

C. Color $\frac{2}{3}$

Write the fraction for the part that is shaded.

D.

$\dfrac{\square}{\square}$

E.

$\dfrac{\square}{\square}$

F.

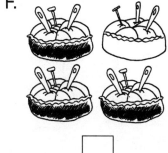

$\dfrac{\square}{\square}$

G. Rina makes a jeweled belt. Color $\frac{3}{10}$ of the jewels red. Color $\frac{4}{10}$ of the jewels blue. Color $\frac{2}{10}$ of the jewels yellow.

H. What fraction of the jewels are not colored? _____

Hare and Tortoise

Compare the fractions. Write >, <, or = in each circle.

A. 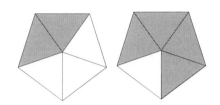   B. 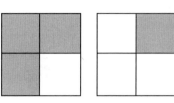   C.

$\frac{2}{5}$ ◯ $\frac{4}{5}$ $\frac{3}{4}$ ◯ $\frac{1}{4}$ $\frac{3}{6}$ ◯ $\frac{3}{6}$

D. $\frac{1}{3}$ ◯ $\frac{2}{3}$ $\frac{1}{5}$ ◯ $\frac{3}{5}$ $\frac{5}{8}$ ◯ $\frac{7}{8}$ $\frac{2}{9}$ ◯ $\frac{1}{9}$

E. $\frac{3}{4}$ ◯ $\frac{2}{4}$ $\frac{1}{2}$ ◯ $\frac{1}{4}$ $\frac{1}{6}$ ◯ $\frac{2}{6}$ $\frac{4}{5}$ ◯ $\frac{4}{5}$

F. $\frac{2}{3}$ ◯ $\frac{1}{3}$ $\frac{5}{6}$ ◯ $\frac{2}{6}$ $\frac{4}{9}$ ◯ $\frac{6}{9}$ $\frac{3}{8}$ ◯ $\frac{6}{8}$

G. A tortoise runs around $\frac{4}{6}$ of the race track. A hare runs around $\frac{7}{9}$ of the race track. Who runs farther? Color the boxes on the tracks to find out.

 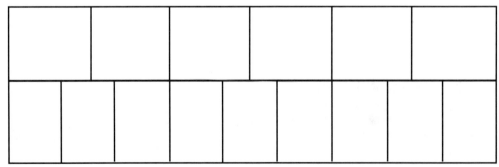

The _____ runs farther.

28   MP3493

Name _____

Cook It Up!

The denominators are the same.
So just add the numerators.

| $\frac{1}{4}$ | $\frac{1}{4}$ | $\frac{1}{4}$ | $\frac{1}{4}$ |

$$\frac{1}{4} + \frac{2}{4} = \frac{3}{4}$$

Color the fractions to add them.

A.

| $\frac{1}{3}$ | $\frac{1}{3}$ | $\frac{1}{3}$ |

$$\frac{1}{3} + \frac{1}{3} = \frac{\square}{\square}$$

B.

| $\frac{1}{6}$ | $\frac{1}{6}$ | $\frac{1}{6}$ | $\frac{1}{6}$ | $\frac{1}{6}$ | $\frac{1}{6}$ |

$$\frac{1}{6} + \frac{3}{6} = \frac{\square}{\square}$$

C.

| $\frac{1}{5}$ | $\frac{1}{5}$ | $\frac{1}{5}$ | $\frac{1}{5}$ | $\frac{1}{5}$ |

$$\frac{2}{5} + \frac{1}{5} = \frac{\square}{\square}$$

D.

| $\frac{1}{8}$ | $\frac{1}{8}$ | $\frac{1}{8}$ | $\frac{1}{8}$ | $\frac{1}{8}$ | $\frac{1}{8}$ | $\frac{1}{8}$ | $\frac{1}{8}$ |

$$\frac{3}{8} + \frac{2}{8} = \frac{\square}{\square}$$

Add the fractions.

E. $\dfrac{1}{8} + \dfrac{5}{8} = \dfrac{\square}{\square}$ $\dfrac{2}{6} + \dfrac{3}{6} = \dfrac{\square}{\square}$ $\dfrac{3}{10} + \dfrac{1}{10} = \dfrac{\square}{\square}$

F. $\dfrac{3}{5} + \dfrac{1}{5} = \dfrac{\square}{\square}$ $\dfrac{1}{4} + \dfrac{1}{4} = \dfrac{\square}{\square}$ $\dfrac{2}{9} + \dfrac{5}{9} = \dfrac{\square}{\square}$

G. Color the brownies to solve the problem. Derek puts red sparkles on $\frac{3}{8}$ of the brownies. He puts green sparkles on $\frac{2}{8}$ of the brownies. What fraction of the brownies have sparkles?

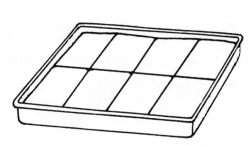

_____

MP3493

Name _____

In the Wood Shop

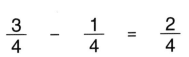

The denominators are the same.
So just subtract the numerators.

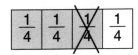

$$\frac{3}{4} - \frac{1}{4} = \frac{2}{4}$$

Color the fractions. Cross out to subtract.

A.

| $\frac{1}{5}$ | $\frac{1}{5}$ | $\frac{1}{5}$ | $\frac{1}{5}$ | $\frac{1}{5}$ |

$$\frac{4}{5} - \frac{1}{5} = \frac{\square}{\square}$$

B.

| $\frac{1}{3}$ | $\frac{1}{3}$ | $\frac{1}{3}$ |

$$\frac{2}{3} - \frac{1}{3} = \frac{\square}{\square}$$

C.

| $\frac{1}{8}$ | $\frac{1}{8}$ | $\frac{1}{8}$ | $\frac{1}{8}$ | $\frac{1}{8}$ | $\frac{1}{8}$ | $\frac{1}{8}$ | $\frac{1}{8}$ |

$$\frac{7}{8} - \frac{5}{8} = \frac{\square}{\square}$$

D.

| $\frac{1}{6}$ | $\frac{1}{6}$ | $\frac{1}{6}$ | $\frac{1}{6}$ | $\frac{1}{6}$ | $\frac{1}{6}$ |

$$\frac{4}{6} - \frac{2}{6} = \frac{\square}{\square}$$

Subtract the fractions.

E. $\dfrac{3}{5} - \dfrac{2}{5} = \dfrac{\square}{\square}$ $\dfrac{5}{6} - \dfrac{1}{6} = \dfrac{\square}{\square}$ $\dfrac{7}{9} - \dfrac{3}{9} = \dfrac{\square}{\square}$

F. $\dfrac{6}{8} - \dfrac{1}{8} = \dfrac{\square}{\square}$ $\dfrac{2}{4} - \dfrac{1}{4} = \dfrac{\square}{\square}$ $\dfrac{9}{10} - \dfrac{5}{10} = \dfrac{\square}{\square}$

G. Draw a picture to solve the problem. Mandy makes a birdhouse with a board of wood. She cuts off $\frac{2}{5}$ of the board. What fraction of the board is left? _____

 MP3493

Name _____

Sewing Class

Add.

A. $\dfrac{1}{4}$ + $\dfrac{3}{4}$ = $\dfrac{\square}{\square}$ $\dfrac{1}{5}$ + $\dfrac{1}{5}$ = $\dfrac{\square}{\square}$ $\dfrac{5}{9}$ + $\dfrac{2}{9}$ = $\dfrac{\square}{\square}$

B. $\dfrac{3}{6}$ + $\dfrac{1}{6}$ = $\dfrac{\square}{\square}$ $\dfrac{1}{3}$ + $\dfrac{2}{3}$ = $\dfrac{\square}{\square}$ $\dfrac{2}{8}$ + $\dfrac{5}{8}$ = $\dfrac{\square}{\square}$

C. $\dfrac{4}{7}$ + $\dfrac{2}{7}$ = $\dfrac{\square}{\square}$ $\dfrac{2}{6}$ + $\dfrac{1}{6}$ = $\dfrac{\square}{\square}$ $\dfrac{2}{10}$ + $\dfrac{6}{10}$ = $\dfrac{\square}{\square}$

Subtract.

D. $\dfrac{4}{5}$ − $\dfrac{3}{5}$ = $\dfrac{\square}{\square}$ $\dfrac{5}{6}$ − $\dfrac{2}{6}$ = $\dfrac{\square}{\square}$ $\dfrac{7}{8}$ − $\dfrac{1}{8}$ = $\dfrac{\square}{\square}$

E. $\dfrac{6}{7}$ − $\dfrac{3}{7}$ = $\dfrac{\square}{\square}$ $\dfrac{3}{4}$ − $\dfrac{1}{4}$ = $\dfrac{\square}{\square}$ $\dfrac{5}{9}$ − $\dfrac{2}{9}$ = $\dfrac{\square}{\square}$

F. $\dfrac{2}{3}$ − $\dfrac{1}{3}$ = $\dfrac{\square}{\square}$ $\dfrac{7}{10}$ − $\dfrac{2}{10}$ = $\dfrac{\square}{\square}$ $\dfrac{4}{8}$ − $\dfrac{3}{8}$ = $\dfrac{\square}{\square}$

G.   Elena sews a ribbon across the bottom of a scarf that is one yard long. She has sewed $\dfrac{3}{8}$ yard of ribbon so far. How much more ribbon does she need to sew to finish the whole scarf?

_____

H.   Ryan cuts $\dfrac{4}{10}$ meter of red fabric to make a kite. He cuts $\dfrac{3}{10}$ meter of blue fabric. He needs $\dfrac{8}{10}$ of a meter to make the kite. Does he have enough fabric? Explain.

_____

Name _____

Band Practice

There are 23 marchers in each row of the band. There are 6 rows of marchers. How many marchers are there in all? 23 x 6	Multiply the tens. Regroup the ones. 1 23 x 6 8 18 ones = 1 ten 8 ones	Multiply the ones. Add the regrouped ten. 1 23 x 6 138 138 marchers in all

Multiply.

A.
32	18	12	41	50	21
x 8	x 3	x 4	x 3	x 7	x 5

B.
23	62	34	72	51	45
x 3	x 4	x 7	x 3	x 4	x 7

C.
83	74	37	94	19	61
x 3	x 6	x 8	x 5	x 7	x 9

D.   Help the marcher find the tuba. Draw a path through the boxes with numbers that match your answers above.

42	231	69	495	350	248	133	305
315	296	115	470	612	246	424	216

MP3493

Name _____

Football Fun

Multiply the ones. Regroup the ones.	Multiply the tens. Add the regrouped ones.	Multiply the hundreds. Add the regrouped hundred.
$\overset{1}{2}45$ x 3 ——— 5	$\overset{11}{2}45$ x 3 ——— 35	$\overset{11}{2}45$ x 3 ——— 735
15 ones = 1 ten 5 ones	13 tens = 1 hundred 3 tens	3 x 245 = 735

Multiply.

A.
```
  213        314        561        174        211
x   3      x   4      x   2      x   5      x   5
———        ———        ———        ———        ———

  F          M          A          B          L
```

B.
```
  304        418        632        369        414
x   2      x   6      x   4      x   5      x   3
———        ———        ———        ———        ———

  H          T          C          S          K
```

C.
```
  842        537        153        227        720
x   4      x   3      x   5      x   6      x   4
———        ———        ———        ———        ———

  U          M          A          R          D
```

D. What football player uses fractions to help him play? To find out, match the letters from your answers above to the numbers below the lines.

A
```
  ____    ____    ____    ____    ____    ____    ____    ____
  608     765    1,055    639     870    1,122   2,528   1,242
```

It's Collectible!

Multiply.

A.   27   44   36   42   98   56
       x 3   x 6   x 4   x 5   x 7   x 6

B.   29   33   72   61   85   89
       x 6   x 3   x 7   x 8   x 4   x 9

C.   132   413   270   561   718
     x 3   x 5   x 2   x 8   x 6

D.   392   528   178   613   357
     x 9   x 4   x 3   x 9   x 7

Use multiplication to solve each problem.

E.   Kaleesha collects baseball cards. She has a book with 135 pages. Each page holds 6 cards. How many baseball cards does Kaleesha need to fill her book?

_____
x _____

_____ cards

F.   Alex collects stamps. He has a book with 218 pages. Each page holds 8 stamps. How many stamps does Alex need to fill his book?

_____
x _____

_____ stamps

 MP3493

At the Ice Cream Parlor

35 scouts go to the ice cream parlor. They sit at tables of 4. How many tables do they need?

Multiply.

$$\begin{array}{r} 8 \\ 4\overline{)35} \\ 32 \end{array}$$

Subtract.

$$\begin{array}{r} 8 \\ 4\overline{)35} \\ -32 \\ \hline 3 \end{array}$$

Write the remainder.

$$\begin{array}{r} 8\ R3 \\ 4\overline{)35} \\ -32 \\ \hline 3 \end{array}$$

The scouts need 9 tables.

Divide.

A. $3\overline{)29}$ $6\overline{)57}$ $5\overline{)44}$ $2\overline{)19}$ $7\overline{)38}$

       A   **L**   **N**   **M**   **E**

B. $4\overline{)33}$ $7\overline{)49}$ $5\overline{)39}$ $7\overline{)31}$ $8\overline{)55}$

       B   **D**   **G**   **I**   **A**

C. $5\overline{)38}$ $4\overline{)26}$ $7\overline{)69}$ $8\overline{)37}$ $6\overline{)51}$

       N   **A**   **S**   **P**   **T**

D. What kind of ice cream sundae likes to divide? To find out, use letters from your answers above.

A ____ ____ ____ ____ ____ ____
   8 R1   9 R2   8 R4   6 R7   7 R3   6 R2

   ____ ____ ____ ____ ____
   9 R6   4 R5   9 R3   4 R3   8 R3

It's Puzzling

Divide the tens. Multiply.	Subtract 4 tens from 47.	Divide the ones. Multiply.	Subtract 4 ones. Write the remainder.

```
      1              1             11           11 R3
  4) 47          4) 47         4) 47         4) 47
  - 40           - 40          - 40          - 40
  ____           ____          ____          ____
                    7             7             7
                                - 4           - 4
                                ____          ____
                                               3
```

Divide.

A.   3) 96   4) 48   5) 65   2) 78   6) 72

B.   4) 85   5) 93   4) 67   7) 98   2) 39

C.   3) 68   3) 74   6) 81   4) 97   2) 76

D.   Color the puzzle pieces with quotients that match your answers above.

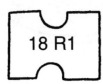
18 R1

16 R3

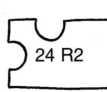
21 R4

24 R2

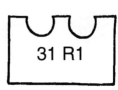
31 R1

MP3493

Pancakes a Plenty

A chef makes 356 pancakes. He puts 3 pancakes on a plate. How many plates can he make?

First, divide the hundreds.

Then divide the tens.

Last, divide the ones.
Write the remainder.

$$
\begin{array}{r}
118 \text{ R2} \\
3\overline{)356} \\
-300 \\
\hline
56 \\
-30 \\
\hline
26 \\
-24 \\
\hline
2
\end{array}
$$

Help the chef find his way to the stove. Divide. Color the chef's caps that have no remainders. Then follow the path.

A. $4\overline{)484}$ $5\overline{)750}$ $3\overline{)654}$ $2\overline{)829}$ $3\overline{)418}$

B. $7\overline{)814}$ $2\overline{)931}$ $4\overline{)672}$ $6\overline{)798}$ $2\overline{)694}$

C. $4\overline{)506}$ $3\overline{)965}$ $8\overline{)819}$ $3\overline{)799}$ $3\overline{)648}$

D. Write a division problem that has 3 digits in the quotient. Give it to a friend to solve.

MP3493

Name _____

Summer Vacation

Solve each word problem.

A. Beth and Jill go to summer camp. There are 9 tents at the campsite. 12 girls stay in each tent. How many girls are at camp?

_____

B. Mr. and Mrs. Perez go on a train trip. Each car on the train holds 54 passengers. There are 8 cars. How many passengers will the train hold?

_____

C. The dining room in a large hotel holds 318 people. If each table seats 6 people, how many tables are needed to fill the dining room?

_____

D. 68 people go on a donkey ride down the Grand Canyon. There are 4 groups in all. How many people are in each group?

_____

E. An airplane holds 285 passengers. The rows of the airplane seat 2 and 3 passengers across. How many rows of seats are in the airplane?

_____

F. Joseph and Malik go to an Action Park. The giant roller coaster holds 153 people. If all the seats are filled, how many people will have ridden the roller coaster after 8 trips?

_____

G. Write and solve your own multiplication or division problem about a summer vacation. Use the box at right.

Name _____

Paint Job

A decimal shows tenths of a fraction.

$\frac{3}{10} = 0.3$

A decimal shows hundredths of a fraction.

$\frac{35}{100} = 0.35$

Write a decimal for the shaded part.

A.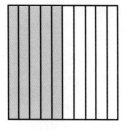

_____

B.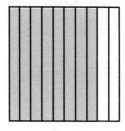

_____

C.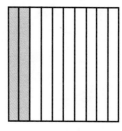

_____

D.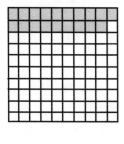

_____

E.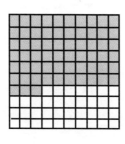

_____

F.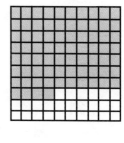

_____

Write each fraction as a decimal.

G. $\frac{4}{10}$ _____ $\frac{9}{10}$ _____ $\frac{1}{10}$ _____ $\frac{7}{10}$ _____ $\frac{6}{10}$ _____

H. $\frac{15}{100}$ _____ $\frac{52}{100}$ _____ $\frac{88}{100}$ _____ $\frac{60}{100}$ _____

I. Tom is painting a fence. Write a decimal to show how much of the fence he has painted so far.

_____

         MP3493

Name _____

Saving Money

Line up the decimal points. Add decimals from right to left.
Write the decimal point and dollar sign in the sum.

6.4	5.47	$ 7.48
+ 1.7	+ 3.98	+ 3.98
8.1	9.45	$11.46

Add.

A.

4.5	5.8	3.9	7.4	6.1
+ 2.3	+ 1.7	+ 4.2	+ 5.9	+ 4.2

B.

3.56	6.09	1.93	5.84	4.67
+ 4.21	+ 2.55	+ 5.72	+ 7.72	+ 8.68

C.

$3.56	$2.89	$5.67	$6.22	$3.64
+ 4.12	+ 1.17	+ 2.83	+ 8.94	+ 9.78

D.

$23.21	$34.67	$52.19	$60.47	$29.67
+ 15.36	+ 12.58	+ 38.40	+ 9.48	+ 85.14

E. Gail has a one-dollar bill, a five-dollar
    bill, and these coins in her piggy bank.
    How much money does she have?

MP3493

Name _____

Spending Money

Line up the decimal points. Subtract decimals from right to left.

7.2	8.16	$ 6.83
− 4.8	− 3.45	− 2.19
2.4	4.71	$ 4.64

Subtract. Remember to write the decimal point and dollar sign in the difference.

A.
7.8	9.2	6.7	8.3	7.7
− 1.5	− 3.6	− 2.1	− 4.9	− 4.8

B.
6.25	8.39	7.35	6.05	5.11
− 2.14	− 4.16	− 1.97	− 4.36	− 4.88

C.
$6.48	$9.00	$5.37	$8.51	$7.08
− 1.36	− 2.83	− 1.54	− 2.96	− 6.53

D.
$45.39	$87.45	$67.45	$83.70	$59.02
− 14.05	− 13.83	− 32.77	− 54.71	− 49.54

E.   Rick has a one-dollar bill, a five-dollar bill, and these coins in his wallet. He spends $5.35. How much money does he have left?

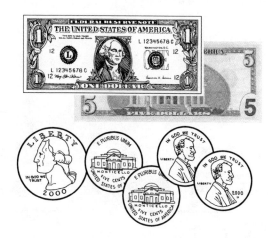

Name _____

Date _____ Score _____

Shade in the circle of the correct answer.

Add or subtract.

1.
$$23$$
$$+ 45$$

○   ○   ○
69   68   58

2.
$$56$$
$$+ 19$$

○   ○   ○
65   75   79

3.
$$73$$
$$- 29$$

○   ○   ○
55   54   44

4.
$$535$$
$$+ 186$$

○   ○   ○
611   621   721

5.
$$304$$
$$- 52$$

○   ○   ○
252   352   452

6.
$$681$$
$$- 196$$

○   ○   ○
485   585   495

7.
$$1,368$$
$$+ 6,947$$

○   ○   ○
8,315   7,215   8,415

8.
$$9,056$$
$$- 4,189$$

○   ○   ○
4,867   4,967   5,977

9.
$$4.9$$
$$+ 5.8$$

○   ○   ○
9.7   10.7   10.8

10.
$$6.4$$
$$- 1.8$$

○   ○   ○
4.5   4.6   5.6

11.
$$\$2.78$$
$$+ 4.51$$

○   ○   ○
$6.19 $7.19 $7.29

12.
$$\$8.23$$
$$- 2.19$$

○   ○   ○
$5.04 $6.04 $6.14

MP3493

Name _____

Date _____ Score _____

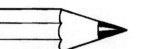

Shade in the circle of the correct answer.

Multiply or divide.

1. $4 \times 9 =$ ____

   ○   ○   ○
   28   32   36

2. $6 \times 7 =$ ____

   ○   ○   ○
   36   42   48

3. $56 \div 7 =$ ____

   ○   ○   ○
   9   8   7

---

4.   5
      x 8
      ─────
   ○ 13
   ○ 40
   ○ 45

5.   6
      x 9
      ─────
   ○ 15
   ○ 48
   ○ 54

6.   7
      x 4
      ─────
   ○ 28
   ○ 11
   ○ 24

---

7. 8)64
   ○ 7
   ○ 8
   ○ 9

8. 6)30
   ○ 6
   ○ 5
   ○ 7

9. 9)27
   ○ 5
   ○ 4
   ○ 3

---

10.   57
       x 8
       ─────

   ○   ○   ○
   456   406   454

11.   368
       x   4
       ─────

   ○   ○   ○
   1,272   1,444   1,472

12.   725
       x   3
       ─────

   ○   ○   ○
   2,175   2,165   2,475

---

13. 7)48
   ○ 7 R6
   ○ 6 R6
   ○   7

14. 4)95
   ○ 23 R1
   ○ 23 R3
   ○ 22 R2

15. 3)815
   ○ 271 R2
   ○ 270 R2
   ○   271

43

MP3493

Name _____

Date _____ Score _____

Shade in the circle of the correct answer.

Identify the shape.

1.

    ○ rectangle

    ○ triangle

    ○ pentagon

_____

2.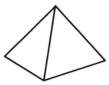

    ○ cube

    ○ cylinder

    ○ pyramid

_____

3.

    ○ cone

    ○ cylinder

    ○ sphere

_____

4. How many faces?

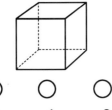

  ○   ○   ○

  3   4   6

_____

5. How many edges?

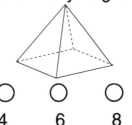

  ○   ○   ○

  4   6   8

_____

6. How many vertices?

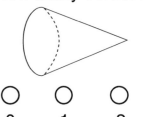

  ○   ○   ○

  0   1   2

_____

Which fraction shows the shaded region?

7.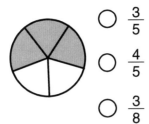

  ○ $\frac{3}{5}$

  ○ $\frac{4}{5}$

  ○ $\frac{3}{8}$

_____

8.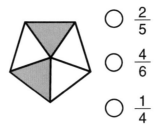

  ○ $\frac{2}{5}$

  ○ $\frac{4}{6}$

  ○ $\frac{1}{4}$

_____

9. Compare

$\frac{4}{5}$ ○ $\frac{1}{5}$

  ○ >

  ○ <

  ○ =

_____

Add or subtract.

10. $\frac{3}{9} + \frac{2}{9} =$ _____

  ○   ○   ○

  $\frac{1}{9}$ $\frac{5}{9}$ $\frac{4}{9}$

11. $\frac{5}{8} - \frac{1}{8} =$ _____

  ○   ○   ○

  $\frac{4}{8}$ $\frac{6}{8}$ $\frac{1}{4}$

12. $\frac{1}{5} + \frac{3}{5} =$ _____

  ○   ○   ○

  $\frac{2}{5}$ $\frac{4}{5}$ $\frac{5}{5}$

Name _____

Date _____ Score _____

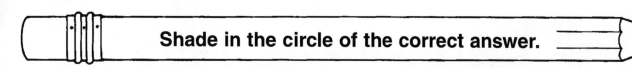

Shade in the circle of the correct answer.

Use the graph to answer questions 1 and 2.

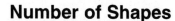

Number of Shapes

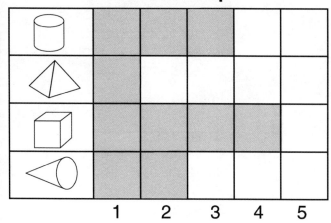

| | 1 | 2 | 3 | 4 | 5 |

Solve each word problem.

3.   On Tuesday, Janine drove 139 miles. On Wednesday, she drove 274 miles. How many miles did Janine drive in all?

○   ○   ○

413   313   303

5.   A toy store gets a shipment of 352 dolls. There are 8 shelves for the dolls in the store. How many dolls can fit on each shelf?

○   ○   ○

42   44   48

1.   How many more cubes are there than pyramids?

○   ○   ○

1   2   3

2.   How many shapes are there in all?

○   ○   ○

8   9   10

4.   A book company packs 8 books in a carton. A trucking company can carry 256 cartons on a truck. How many books in all can fit on the truck?

○   ○   ○

1,648   2,008   2,048

6.   Greenville is 734 miles from Lansing. Park City is 478 miles from Lansing. How many more miles from Lansing is Greenville than Park City?

○   ○   ○

1,212   356   256

                                       MP3493

Answer Key

Page 1
A. 564
B. 329
C. 803
D. 270
E. 3,718
F. 5,825
G. 4,052
H. 9,409
I. 81, 82, 85
J. 439, 440, 444
K. 2,112; 2,113; 2,115
L. 335, 340, 345
M. 4,566; 4,568; 4,570
N. 593
O. 124

Page 2
A. 8, 13, 13, 10, 5, 14, 9
B. 6, 4, 3, 8, 6, 6, 8
C. 7, 14, 14, 12, 12, 11, 13
D. 8, 8, 3, 9, 9, 8, 9
E. 12, 9, 6, 2, 10, 7, 17
F. 7, 1, 18, 10, 15, 7, 16

Page 3
A. 8, 5, 8, 13
B. 16, 7, 16, 7
C. 9, 9, 10, 19
D. 3, 3, 11, 3
E. 9, 9, 6, 15
F. 7, 7, 7, 17
G. 8, 8, 12, 4
H. 14, 8, 6, 14
I. 6, 6, 6, 13

Page 4
A. 37, 56, 68, 78, 75, 37
B. 73, 89, 58, 46, 62, 18
C. 56, 86, 94, 65, 77, 52
D. 70, 63, 117, 131, 54, 156
E. 128, 149, 136, 125, 110, 167
F. 52, 77, 56, 65

Page 5
A. 33, 32, 42, 64, 51, 40
B. 15, 52, 39, 64, 68, 21
C. 48, 17, 43, 23, 65, 8
D. 71, 38, 9, 48, 32, 72
E. 19, 58, 30, 26, 18, 17

F. 5, 2, 3; 4, 1, 3
G. 6, 1, 5; 6, 1, 5
H. 5, 5, 0; 1, 1, 0

Page 6
A. 771, 399, 537, 821, 1,281
B. 881, 563, 379, 467, 871
C. 525, 818, 493, 775, 993
D. 733, 423, 610, 823, 641
E. 861, 1,094, 840, 1,348, 1,335
F. 165, 135
G. 270, 130

Page 7
A. 511, 712, 272, 811, 321
B. 338, 527, 142, 433, 857
C. 657, 455, 85, 292, 685
D. 488, 258, 578, 868, 479
E. 701, 287, 265, 386, 655
F. 5, 2, 1
G. 3, 4, 8
H. 6, 3, 1

Page 8
A. 92, 54, 112, 90, 27
B. 133, 530, 713, 921, 639
C. 577, 905, 282, 234, 989
D. 114
E. 79
F. 40, 37

Page 9
A. 4,039; 6,932; 6,767; 5,714; 4,237
B. 7,937; 1,878; 6,414; 7,367; 9,569
C. 2,553; 5,712; 8,321; 45,714; 64,237
D. 16,010; 77,582; 76,908; 36,447; 94,237

Answer to riddle: A BOX OF QUACKERS

Page 10
A. 2,120; 4,933; 2,612; 6,549; 5,482
B. 4,099; 1,713; 5,457; 3,005; 919
C. 1,188; 386, 6,454; 29,679; 50,959
D. 38,424; 45,466; 6,424; 52,325; 76,731
E. Magic number: 3
F. Magic number: 5

Page 11
A. 907 points; yes
B. 7,860 points; yes
C. subtract; 1,797 more points
D. Jamal and Lin; 98,006
E. 19,967 points
F. 55,847 points; yes

Page 12
A. 25, 4, 8
B. 16, 5, 45
C. 12, 15, 10, 20, 18, 30, 6
D. 10, 14, 40, 35, 2, 15, 45
E. 16, 30, 18, 12, 0, 20, 14
F. 5 x 5 = 25

Page 13
A. 3, 6, 28
B. 20, 9, 32
C. 24, 15, 8, 6, 8, 4, 21
D. 12, 28, 0, 27, 18, 12, 15
E. 24, 27, 36, 24, 24, 16, 18
F. 4 x 8 = 32

Page 14
A. 42, 49, 48
B. 54, 35, 56
C. 63, 36, 63, 40, 45, 42, 24
D. 28, 64, 48, 72, 81, 54, 56
E. 7; 8; 8 x 8 = 64

Page 15
A. 42, 20, 27, 8, 12, 14, 25
B. 21, 64, 35, 18, 24, 24, 63
C. 30, 16, 54, 40, 9, 48, 49
D. 81, 28, 56, 18, 32, 15, 36
E. A FACTOR TREE
F. 3, 2, 5, 7

Page 16
A. 4, 2, 6, 2
B. 8, 9, 6, 9
C. 3, 8, 6, 7, 3
D. 2, 7, 4, 1, 5
E. 5, 1, 8, 10, 7
F. 10 times

Page 17
A. 3, 1, 7, 8
B. 3, 2, 8, 5

Answer Key

C. 6, 7, 4, 1, 6
D. 9, 10, 9, 3, 5
E. 2, 5, 10, 4, 7
F. 8 times

Page 18
A. 4, 7, 8, 5
B. 8, 7, 9, 7
C. 5, 9, 8, 9, 6
D. 7, 9, 6, 3, 4
E. 6, 6, 8, 5, 4

Page 19
A. 8, 2, 2, 6, 4, 7, 9
B. 8, 5, 3, 7, 3, 6, 8
C. 8, 9, 6, 3, 4, 1, 7
D. 4, 7, 4, 3, 8, 9, 8
E. A GROUPER FISH
F. 24, 12, 18, 30

Page 20
A. 7, 0, 8
B. 9, 0, 0
C. 0, 6, 7, 0, 6, 5, 0
D. 6, 0, 4, 0
E. 0, 9, 8, 1
F. 4, 7, 0, 0, 3
G. 8, 0, 0, 2

Page 21
A. 27, 12, 28, 16, 30, 35, 9
B. 63, 21, 16, 56, 48, 12, 25
C. 24, 24, 45, 32, 49, 81, 54
D. 5, 4, 8, 2, 3
E. 6, 7, 5, 0, 3
F. 2, 6, 5, 1, 6
G. Car B finishes first.

Page 22
A. 8 x 7 = 56 cupcakes
B. 45 ÷ 9 = 5 riders
C. 36 ÷ 6 = 6 balls
D. 35 ÷ 5 = 7 people
E. 5 x 4 = 20 pigs
F. 5 x 6 = 30 tickets
G. 8 x 4 = 32 pies
H. 7 x 4 = 28 chickens

Page 23
A. triangles: 3, quadrilaterals: 4, pentagons: 2, hexagons: 1
B. 3 + 1 = 4
C. 4 + 2 = 6

Page 24
A. cubes: 3, cones: 2, cylinders: 3, spheres: 1, pyramids: 4
B. 2 + 3 + 4 = 9
C. 1 + 3 = 4

Page 25
A. square; 6, 12, 8
B. triangle, square; 5, 8, 5
C. circle; 1, 0, 1
D. circle; 2, 0, 0
E. none; 0, 0, 0
F. cube and pyramid

Page 26
A. Check students' drawings.
B. Check students' drawings.
C. Check students' drawings.
D. 1/2
E. 2/3
F. 4/6
G. Check students' drawings.
H. Check students' drawings.

Page 27
A. Check students' drawings.
B. Check students' drawings.
C. Check students' drawings.
D. $\dfrac{1}{2}$
E. $\dfrac{2}{8}$
F. $\dfrac{3}{4}$
G. Check students' drawings.
H. $\dfrac{1}{10}$

Page 28
A. <
B. >
C. =
D. <, <, <, >
E. >, >, <, =
F. >, >, <, <
G. The hare runs farther.

Page 29
A. $\dfrac{2}{3}$
B. $\dfrac{4}{6}$
C. $\dfrac{3}{5}$
D. $\dfrac{5}{8}$
E. $\dfrac{6}{8}$, $\dfrac{5}{6}$, $\dfrac{4}{10}$
F. $\dfrac{4}{5}$, $\dfrac{2}{4}$, $\dfrac{7}{9}$
G. $\dfrac{5}{8}$

Page 30
A. $\dfrac{3}{5}$
B. $\dfrac{1}{3}$
C. $\dfrac{2}{8}$
D. $\dfrac{2}{6}$
E. $\dfrac{1}{5}$, $\dfrac{4}{6}$, $\dfrac{4}{9}$
F. $\dfrac{5}{8}$, $\dfrac{1}{4}$, $\dfrac{4}{10}$
G. $\dfrac{3}{5}$

Page 31
A. $\dfrac{4}{4}$, $\dfrac{2}{5}$, $\dfrac{7}{9}$
B. $\dfrac{4}{6}$, $\dfrac{3}{3}$, $\dfrac{7}{8}$
C. $\dfrac{6}{7}$, $\dfrac{3}{6}$, $\dfrac{8}{10}$

Answer Key

D. $\frac{1}{5}$, $\frac{3}{6}$, $\frac{6}{8}$

E. $\frac{3}{7}$, $\frac{2}{4}$, $\frac{3}{9}$

F. $\frac{1}{3}$, $\frac{5}{10}$, $\frac{1}{8}$

G. $\frac{5}{8}$

H. No, he needs another $\frac{1}{10}$ meter.

Page 32
A. 256; 54; 48; 123; 350, 105
B. 69, 248, 238, 216, 204, 315
C. 249, 444, 296, 470, 133, 549
D. Path: 315, 296, 69, 470, 350, 248, 133, 216

Page 33
A. 639; 1,256; 1,122; 870; 1,055
B. 608; 2,508; 2,528; 1,845; 1,242
C. 3,368; 1,611; 765; 1,362; 2,880
D. A HALFBACK

Page 34
A. 81, 264, 144, 210, 686, 336
B. 174, 99, 504, 488, 340, 801
C. 396; 2,065; 540; 4,488; 4,308
D. 3,528; 2,112; 534; 5,517; 2,499
E. 810 cards
F. 1,744 stamps

Page 35
A. 9 R2, 9 R3, 8 R4, 9 R1, 5 R3
B. 8 R1, 7, 7 R4, 4 R3, 6 R7
C. 7 R3, 6 R2, 9 R6, 4 R5, 8 R3
D. A BANANA SPLIT

Page 36
A. 32, 12, 13, 39, 12
B. 21 R1, 18 R3, 16 R3, 14, 19 R1
C. 22 R2, 24 R2, 13 R3, 24 R1, 38
D. 16 R3 and 24 R2 match

Page 37
A. 121, 150, 218, 414 R1, 139 R1
B. 116 R2, 465 R1, 168, 133, 347

C. 126 R2, 321 R2, 102 R3, 266 R1, 216
   Path: 150, 218, 168, 133, 347, 216
D. Check students' problems.

Page 38
A. 108 girls
B. 432 passengers
C. 53 tables
D. 17 people
E. 57 rows
F. 1,224 people
G. Check students' problems.

Page 39
A. 0.5
B. 0.8
C. 0.2
D. 0.19
E. 0.63
F. 0.74
G. 0.4; 0.9; 0.1; 0.7; 0.6
H. 0.15; 0.52; 0.88; 0.60
I. 0.3

Page 40
A. 6.8; 7.5; 8.1; 13.3; 10.3
B. 7.77; 8.64; 7.65; 13.56; 13.35
C. $7.68; $4.06; $8.50; $15.16; $13.42
D. $38.57; $47.25; $90.59; $69.95; $114.81
E. $6.46

Page 41
A. 6.3; 5.6; 4.6; 3.4; 2.9
B. 4.11; 4.23; 5.38; 1.69; 0.23
C. $5.12; $6.17; $3.83; $5.55; $0.55
D. $31.34; $73.62; $34.68; $28.99; $9.48
E. $1.02

Page 42
1. 68
2. 75
3. 44
4. 721

5. 252
6. 485
7. 8,315
8. 4,867
9. 10.7
10. 4.6
11. $7.29
12. $6.04

Page 43
1. 36
2. 42
3. 8
4. 40
5. 54
6. 28
7. 8
8. 5
9. 3
10. 456
11. 1,472
12. 2,175
13. 6 R6
14. 23 R3
15. 271 R2

Page 44
1. rectangle
2. pyramid
3. cylinder
4. 6
5. 8
6. 1
7. 3/5
8. 2/5
9. >
10. 5/9
11. 4/8
12. 4/5

Page 45
1. 3
2. 10
3. 413
4. 2,048
5. 44
6. 256

MP3493